景观再生：
滨水门户的活力复兴

Landscape Regeneration:
The Revitalization of the
Waterfront Portal

———

同济大学建筑与城市规划学院
景观学系　　　著

同济大学出版社
Tongji University Press

夏令营组委会

主　席：李振宇
副主席：孙彤宇　李翔宁　韩锋
成　员：王云才　金云峰　董楠楠　王晓庆　唐育红　李薇

执行委员会

王云才　　董楠楠　　戴代新　　陈筝　　沈洁　　杨晨　　马魏魏
翟宇佳　　张琳

指导教师

董楠楠　戴代新（同济大学建筑与城市规划学院景观学系）
Nathan Heavers（美国弗吉尼亚理工大学景观学系）
Jim Ayorekire（乌干达马凯雷雷大学）
Nick Nelson（美国哈佛大学景观学系）
Andrew Saniga（澳大利亚墨尔本大学景观学系）

答辩委员

孙彤宇　王云才（同济大学建筑与城市规划学院）
张雯（上海市宝山区规划和土地管理局）
David Thompson（美国SWA设计集团）

本书美编小组成员（拼音顺序）

胡玎　　金雅萍　　林梦楠　　王玮祎

目录 Contents

"滨水景观再生"同济CAUP国际设计夏令营   1
The Revitalization of the Waterfront Portal: Tongji
CAUP International Design School

夏令营作品
Works

港口实验室   18
The Port Lab

鱼菜共生 与潮为友   34
Aquaponic Aquaponding

交汇   50
Confluence

城市湿地原型   68
Prototyping Urban Wetland

惊奇 韵律 释放   86
Riddle Rhythm Release

吴淞生态结   98
Wusong Eco-Nexus

后记   114
Postscript

"滨水景观再生" 同济CAUP国际设计夏令营

每年一度的国际设计夏令营是同济大学建筑与城市规划学院具有国际影响的传统项目之一,从 2005 年始办,由建筑、城市规划、景观学三系轮流承办,到 2017 年由景观学系承办时是第十三届。

本届夏令营以"滨水景观再生"为主题,并契合上海黄浦江沿江绿道贯通、还江于民的大背景。夏令营汇集了来自全球 9 个国家、24 所高校的师生共同参与。36 名学生分成 6 个工作团队,在现状踏勘的基础上,对吴淞炮台湾湿地森林公园南入口区域地块提出了各具特色的发展策略和设计方案。6 个工作团队的设计主题分别为"港口实验室""鱼菜共生 与潮为友""交汇""城市湿地原型""惊奇 韵律 释放"以及"吴淞生态结"。本书呈现的就是这 6 个脚踏实地又充满想象力的方案。

本书对于探讨"滨水景观再生"设计概念、畅想大都市滨水景观面貌,具有一定抛砖引玉的价值。

The Revitalization of the Waterfront Portal: Tongji CAUP International Design School

Tongji CAUP International Design Summer School, owning international influence, is a traditional annual event hosted by three departments of Architecture, Urban Planning and Landscape in turn. Started from 2005, it had been successfully held 13 times. The Design Summer School 2017 was hosted by the Department of Landscape.

This year, the event is themed as The Revitalization of the Waterfront Portal, which is inspired by the connecting of waterfront green land along the Huangpu River and the political direction of returning the river back to people. The event assembles 6 professors and 36 students coming from 9 countries and 24 universities and divided into 6 work teams. Based on fieldwork, each team develops their unique developing strategy and design plan. The 6 grounded and imaginative design themes are The Port Lab, Aquaponic Aquaponding, Confluence, Prototyping Urban Wetland, Riddle Rhythm Release and Wusong Eco-Nexus. All of them will be represented in this book.

Freely thinking about the future and initiating further research of the revitalization of the waterfront portal in big cities is the very value of the book.

主题

水者,万物之本源也。滨水空间兼具自然与人工特征,是城市中不可多得的宝贵资源,它是"人们逃离拥挤的、压力锅式的城市生活的机会,是人们在城市生活中获得呼吸清新空气的疆界的机会"(亚瑟·柯顿·摩尔)。自20世纪70年代以来,滨水空间的再开发利用更被视为城市发展与转型的契机与途径。

虽然滨水空间的改造更新已经被不同的城市反复地讨论与实践,本次设计夏令营仍希望借由水滨门户这样一个特殊的选点,去探讨基于本土的地标性滨水空间的复兴模式。如何通过景观再生的手段,去缝合城市与水滨之间的割裂,让滨水空间回归于市民?如何为衰落的城市水岸注入活力,实现历史文脉的延续、生态环境的修复、城市公共空间品质与城市形象的提升?更为重要的是,如何通过我们大胆的想象与务实的态度,表达出具有上海特质的活力水滨的设计意象?

希望通过本次国际设计夏令营,加强人居环境设计的国际学术交流,展现独创性的优秀设计思考和方案,并推进这一领域优秀人才的培养。

背景

上海——中国当代最大的金融中心与贸易港口,在19世纪由于良好的港口位置开始展露锋芒,成为中国近代工业的发祥地,以及中国乃至远东地区重要的工业基地。在长江与黄浦江流经的区域,岸线一侧聚集的工业生产活动遗迹见证了上海城市发展的辉煌历史,也造成了滨水空间与城市生活的割裂,遗留的工业污染还对城市环境产生了恶劣影响。1990年代以来,随着产业结构的调整,上海城市发展迈入转型升级、存量更新的阶段,两江沿岸原来以工业和码头仓储为主的功能布局已经不再适应城市发展,转型成为必然。

2002年,上海正式启动黄浦江两岸综合开发,规划范围从吴淞口至徐浦大桥,涉及浦东、宝山、杨浦、虹口、黄浦、卢湾和徐汇7个行政区,由此,沿江地区整体发展战略初步形成,黄浦江从传统生产、航运功能向以金融贸易、文化旅游、生态休闲为主的综合服务型功能转变。

区位概况

基地位于上海市宝山区,隶属于黄浦江两岸综合开发中黄浦江北延伸段WN7单元。此地地处长江与黄浦江交汇处的吴淞口,是水路进入上海市区的咽喉要塞,自古就有"水陆要冲,苏松喉吭"之说,水陆交通条件优越。

318国道
318 National Road

长江
The Yangtze River

同济路（上海绕城高速）
Tongji Road (Shanghai Ring Expy)

场地位置
SITE LOCATION

SITE
场地

设计范围

基地位于吴淞口西侧湾区塘后路以东、塘后支路以北、炮台湾湿地森林公园以南，东接吴淞口灯塔与导堤，临江处有近270°江景，视野非常开阔。规划总面积约25公顷，其中陆地面积约14公顷。

基地概况

1. 现状

基地现状为军事用地，内有舰船器材仓库、舰艇岸勤部供应站、训练中心、车库、勘测队等，但地块内部存在大量闲置或出租的现象，军事功能已十分弱化。现状用地以低层建筑为主，质量一般；现状滨水空间缺乏设计，生态景观形象不理想。

2. 已批控规用地情况

规划以公园绿地为主，有一处住宅组团用地和一处特殊用地。

3. 开发动态

已建设地块和可建设范围见右图，已建设地块内的建筑根据方案设计需求可适当进行改造和拆除。

设计目标

本次设计将基地定位为从长江进入上海的第一道滨水城市门户景观，并兼顾吴淞炮台湾湿地森林公园配套服务区功能。旨在探讨适合上海本土的滨水空间的更新与发展模式，以实现城市文脉的延续、生态环境的修复、城市公共空间品质与城市形象的提升，让衰落的滨水空间恢复活力，重获新生。

设计建议内容

关注点1：提升滨江空间品质，塑造特色滨水门户形象

a. 提升公园服务品质

为达到黄浦江沿岸公众性、开放性和连续性的要求，现状用地作为公园绿地及配套服务设施用地，新增滨江绿地广场、游客服务、小型商业服务、停车设施等功能，通过用地功能的优化，提升沿江城市形象。

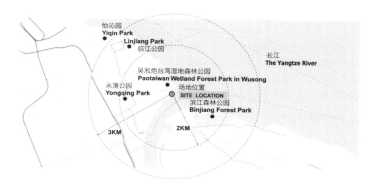

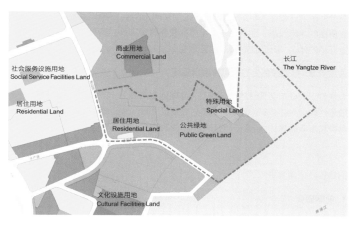

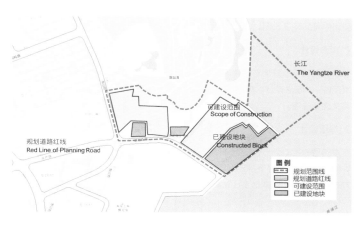

b. 打造特色滨水景观

此地块在控规中的功能为公园绿地(G1)。对现状已有建筑进行分析评估,在以绿化功能为主的基础上,规划建议根据地区发展需求,增加为公园配套服务的公益性设施,完善功能,提升沿江景观环境和城市形象。打造为公众服务的生态滨水景观,营建舒适的人行环境,形成人性化的生态滨水岸线。

c. 文化资源整合

基地内布局有规划展示馆、海军博物馆,也是吴淞开埠的核心地区,具有相对较好的文化资源,可以作为地区公共空间布局的重要节点加以利用。

关注点2:营造舒适而有活力的休闲空间

充分利用滨江资源,营造多层次公共活动空间,结合公园配套服务功能,提供舒适便捷的休闲设施。注重人在环境中的使用感受,提供与滨水空间、自然环境亲近互动的体验,形成活跃、舒适的开放空间。

关注点3:打造绿色生态的可持续示范区

地块位于长江与黄浦江的河口地区,与国际生态岛——崇明岛隔江相望,生态敏感性较强,生物多样性较高。规划应满足生态保育的要求,重点考虑对湿地、鸟类和水质的保护,结合公园打造生态滨江岸线。

最终设计成果要求

1. 前期研究

基于上位规划与基地现状,提出待解决的关键问题及解决问题的方式;对适合上海本土的滨水空间的更新与发展模式进行探讨。成果须包括:分析图、概念图和中英文文字说明。

2. 概念设计

要求在充分考虑基地与周边区域以及更大范围关系的基础上,完成基地(25公顷范围)的概念设计。成果须包括:基地分析图、概念发展图、概念设计图和中英文文字说明。

3. 详细设计

要求选取一至二处节点进行详细设计。设计成果须包括:总平面图、竖向设计图、种植设计图,以及必要的剖面图、透视图、分析图和中英文文字说明等,图纸表现方式以及具体比例依据设计对象的特征确定。

THEME

Water nurtures everything. Waterfront, an urban area of both natural and artificial, is a valuable resource for people to escape from the crowded urban life, and to breathe fresh air, as Arthur Cotton Moore pointed out. Since the 1970s, the regeneration of waterfront has been regarded as an important opportunity for city development and transition.

Although the redevelopment of the waterfront has been discussed and practiced in many cities in the world, in this Summer School we aim to reinvent a waterfront portal, a critical landmark in Shanghai, exploring a localized model for waterfront landmark revitalization. How to bridge the gap between the city and the river through landscape regeneration, returning the waterfront to the citizens? How to revive the waterfront? More importantly, how can we create an active waterfront with Shanghai local features through out-of-box design and sound justifications?

Finally, we hope this summer school may offer an opportunity for international academic communication, a stage to share creative design ideas, and a platform to foster brilliant future designers.

BACKGROUND

Shanghai—the largest financial center and trade port of China today, started to stand out in the 19th century due to its excellent location. Shanghai was the birthplace of Chinese modern industry and an important industrial base of China and the Far East. In the area where the Yangtze River and Huangpu River flow through, industrial production sites are distributed along the river, witnessing the glorious history of Shanghai. However, these sites also created barriers between waterfront spaces and city life, and industrial pollution made negative impacts on the city environment. Since the 1990s, with the adjustment of industrial structure, Shanghai stepped into the stage of transformation and upgrading. The functions of the waterfront spaces, mainly industry and terminal storage, could no longer meet the demand of city development. Thus, a transformation has become an inevitable trend.

In 2002, Shanghai officially launched the general development plan of Huangpu River. Such development has involved 7 districts including Pudong, Baoshan, Yangpu, Hongkou, Huangpu, Luwan and Xuhui, extending from Wusong Port to Xupu Bridge. Therefore, the overall development strategy of the area along the Huangpu River has preliminarily formed—the function of the river has transformed from traditional production and shipping to the service industry such as financial trade, cultural tourism, and ecological recreation.

SITE LOCATION

The site is located in Baoshan District in Shanghai. It is part of the Unit WN7 of the north extension of the development area along the Huangpu River according to the general development plan of the waterfront spaces. The site is located in Wusongkou—by the confluence of the Yangtze River and the Huangpu River, which is a key point to pass when entering the Shanghai City by water. As the old saying goes, Wusongkou is "a transportation hub of land and water, and the thoroughfare of Suzhou and Songjiang (today's Shanghai)", which demonstrates its exceptional advantage in transportation.

DESIGN SCOPE

The site is located in the western bay of Wusongkou, east of Tanghou Road, north of Tanghou Bypass, south of Paotaiwan Wetland Forest Park, and it adjoins the lighthouse and jetty in the east. It enjoys 270° river view. It covers a total area of about 25 hectares, including 14 hectares of land.

SITE INFORMATION

1. Present situation

Presently, the site is a military land, in which there are ship equipment warehouses, a supply station, a training center, garages, survey team offices, etc. But a large part of the parcel is idle or leased, leading to weakened military function. The land is mostly occupied by ordinary low-rise buildings; a lack of waterfront space design leads to poor ecological landscape.

2. Approved regulatory land use

The plan mainly covers land for green belt, and a little for residential or other use.

3. Development status

The figure on p.7 is the built-up parcel and constructible scope.The buildings within the built-up parcel may be transformed and dismantled according to design requirements.

THE GOAL OF DESIGN

In this project, the site should be considered as the portal landscape of Shanghai on the Yangtze River. It should also be considered as a functional area of Paotaiwan Wetland Forest Park. The aim of this design is to explore contextualized renewal and development strategies for waterfront spaces of Shanghai. These new strategies could help us to maintain urban cultures, to restore the ecological environment and to improve public spaces, and finally to revitalize the declining waterfront spaces.

DESIGN PROPOSAL

Focus 1: Improving waterfront space quality, building the characteristic waterfront gateway

a. Improve park service quality

According to the requirement for publicness, openness and continuity on the areas along the Huangpu River, the present land is used as park greenbelt and area for service facilities, and the newly added part includes square, tourist service, small business service, parking facility, etc. The waterfront city image will be improved through land use optimization.

b. Creating distinctive waterfront landscape

The parcel is park greenland (G1) according to the regulatory planning. The design team shall analyze and assess existing buildings, make plans to meet regional development needs in addition to greening function, increase public facilities for park matching service, improve waterfront landscape environment and city image. Build an ecological waterfront landscape for public service, create a comfortable pedestrian environment and form user-friendly ecological waterfront shoreline.

c. Integrating cultural resources

The site comprises a planning exhibition center and a naval museum, and it is also the core area of Wusong Port, so it enjoys good cultural resources. It can also be used as an important node of public space in the area.

Focus 2: Creating a comfortable and vibrant leisure space

Make full use of waterfront resources, create multi-level public space, provide comfortable and user-friendly leisure facilities according to park matching service function. Pay attention to user-friendliness, provide experience in interaction with waterfront space and natural environment, form active and comfortable open space.

Focus 3: Building the demonstration area of sustainable green ecology

The parcel is located in the estuary where the Yangtze River and the Huangpu River meet. It faces international eco-island—Chongming Island across the river, so it is ecologically sensitive and biologically diverse. The plan should meet ecological protection requirement, give priority to protecting wetland, birds and water quality and build ecological waterfront shoreline together with the park.

THE EXPECTED DESIGN OUTCOMES

1. Preliminary Study

Based on the master plan and the condition of the site, students need to identify problems and corresponding strategies to solve them; based on that, students also need to discuss appropriate renewal and development models for waterfront spaces in Shanghai.

The expected outcomes include—site analysis drawings and concept drawings with statements in both Chinese and English.

2. Concept Design

Students are asked to complete the concept design based on a comprehensive understanding of the site (25 hectares) and its context. The outcomes include—site analysis drawings, concept development drawings and conceptual design drawings with statements in both Chinese and English.

3. Detailed Design

Students need to select one or two nodes within the site to do the detailed design. The presentation of drawings and scales will be determined according to the design object. The outcomes may include

—master plan layout, grading plan, planting design, and the necessary sectional drawings, perspectives, analysis drawings with statements in both Chinese and English.

夏令营作品
Works

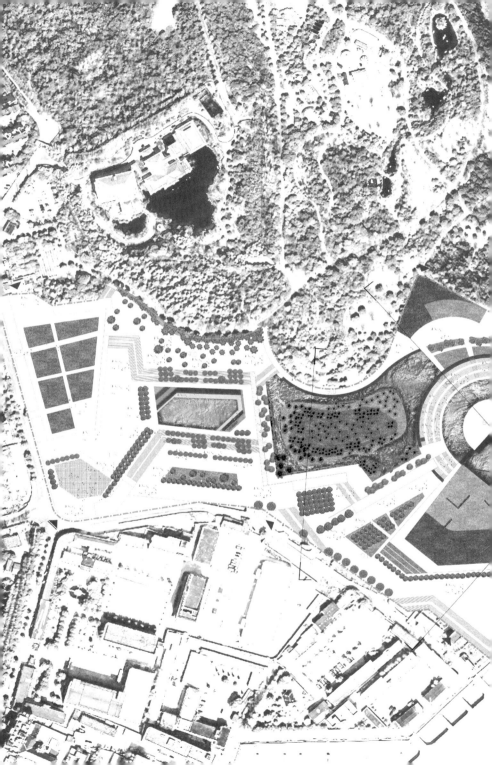

港口实验室
The Port Lab

组员 Members
朱旭东，张莹，Zeynep Goksel,
Zachary Lancaster, 符思熠,
Rena Balfanz

导师 Tutor
Nathan Heavers

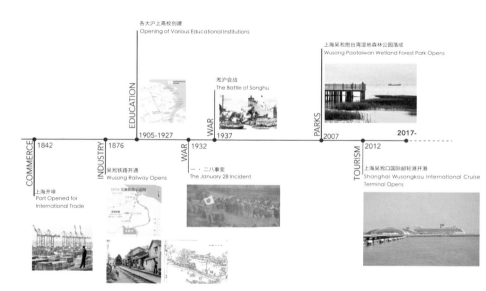

时间轴
Chronology

目前现场是由坚实的硬景材料,社区生活和军事活动的痕迹,以及原生的、未维护的树木和灌木所构成。军事屏障和乱丢的垃圾成为局部点缀。

场地氛围
Site Atmosphere

The current experience of the site is defined by an overbearing presence of hardscape materials and surfaces, traces of community life and military activity, and unmaintained trees and shrubs in their natural state. The site is sprinkled by the military barriers and rubbish littered everywhere.

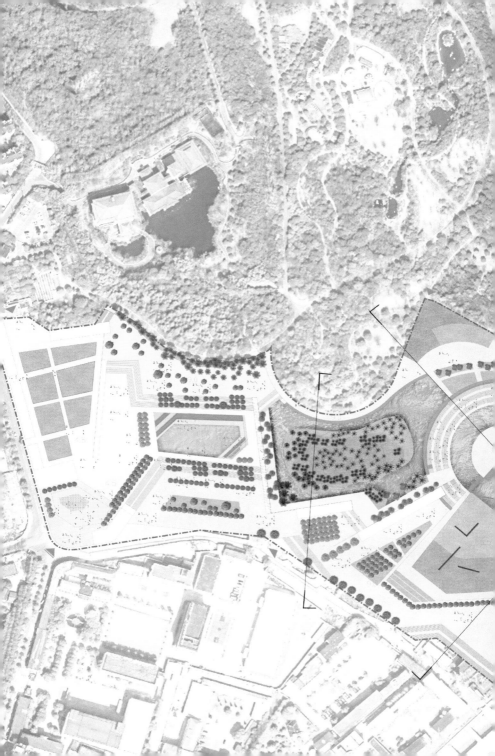

总平面
Masterplan

关键要素
Key Elements

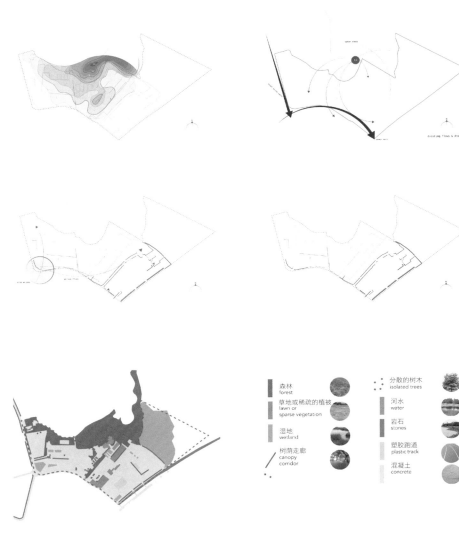

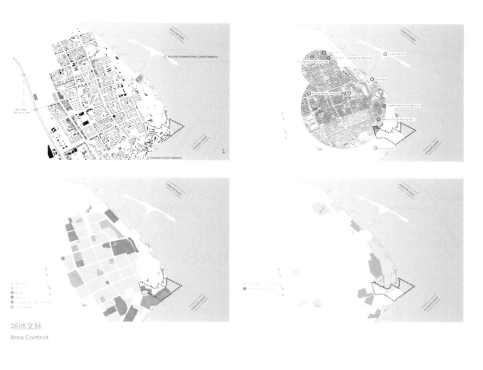

场地文脉
Area Context

作为曾经的军事基地,场地位于长江入海前和黄浦江交汇的地带,也是进入上海的第一道关口。因此,这里汇集着世界上最为活跃的海运港口以及数十条轮运航线与水上线路。

尽管拥有着优越的水上地理条件,陆上的公共交通设置却不尽如人意。离场地最近的地铁站在15分钟的适宜步行路程之外,而乘地铁到外滩路上总共需要花费2个小时左右,即使驾车也需要45分钟以上。虽然在最近的高速公路入口和地铁轨交站旁有一个发展中的商业走廊,场地周边仍以居民区为主,其中最令人瞩目的是各种各样的教育设施以及公共机构。

As a former military base, the site is located at the doorstep of Shanghai and the intersection of the Huangpu River and the Yangtze River. While its waterfront location is ideal, which is home to the most active international cruise terminal in the world and several shipping and ferry routes, land connections are challenging: the closest metro station is more than 15-minute walk from the site which is nearly two hours to the Bund or 45 minutes by car. Despite a growing commercial corridor close to main highways and metro stations, the surroundings are largely residentian area, and feature a diverse distribution of educational facilities and public institutions.

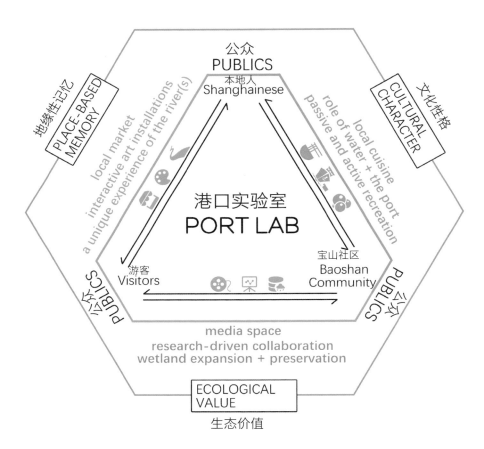

通过本地市场、互动艺术装置以及独特的滨水体验，使人们的身份在本地人和游客之间转换。基于本地航线，水与港口提供一种被动却频繁的日常消遣，使上海市区与宝山社区相互通达。媒体空间在游客与宝山社区之间建立互动，用研究驱动型的合作来拓展和保护湿地。

滨水复兴需要加强场地中历史、文化和物理要素之间的连接，而港口实验室正是强调了在21世纪的上海日益需要的连接、合作和互动。长江入海前和黄浦江的交汇，汇集着国际、上海和区域的影响，赋予了这块场地独一无二的角色，同时也给予场地新的机会，去运用人群、水流和生态效应贯通那些现在隔离它们的墙壁。通过置入一些项目去打开景观空间，利用具有社会意义和空间效应的项目去阐述和加强这块场地内的联结和互动。同时，河流交汇处的水流提供了将人文与水文联结在一起的机会。

基于这块场地的特殊性，港口实验室重新考虑了以空间上的阻隔（墙和堤坝）去创造一个无论是空间层面还是社会层面（包括艺术家、设计师、学者和公众）都开放的空间。这个持续的景观实验向活动在这个场地上的人群提出一个关键的问题——如何去应对和思考城市人群与这块场地上持续变化的景观之间的关系？

Through local markets, interactive art installations and unique experiences of the rivers, the characters of people keep changing between Shanghainese and visitors. According to the local cuisine, the role of water and the port generates a passive but active recreation for Shanghainese in Baoshan Community. On the other hand, a positive interaction is formed among visitors and Baoshan Community by media space, which promotes wetland expansion and preservation based on research-driven collaboration.

For the revitalization of the port, it is necessary to enhance the connection of historical, cultural and physical elements on the site and this is what the Port Lab emphasizes in Shanghai, a modern city expecting connection, cooperation and interaction in the 21st century. The confluence of the Yangtze River and the Huangpu River, where the site is located, assembles multiple power from global, Shanghai and region. It not only grants unique characteristic to the site but also provides various new opportunities to take advantage of people, stream and ecological effect. It is possible to push over the walls now separating them by implanting projects which are full of social meanings and spatial influence. Meanwhile, the stream can also play an important role to reconnect culture and hydrology.

Based on these particularities, the Port Lab in Baoshan is meant to create a space which is open no matter in spatial or social context, including artists, designers, scholars and the public. The persistent landscape experiment raises a question to all the people on the site—how to consider and deal with the relationship between citizens and the constantly changing landscape on the site?

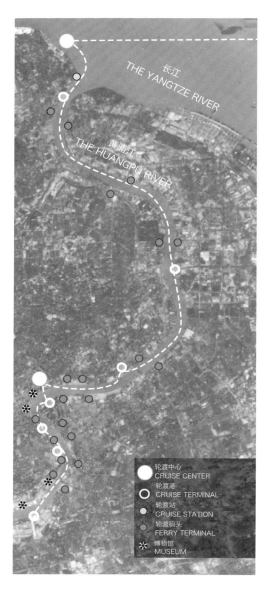

通过对这些不同的公众的认识,实验室被定义为景观设计中的合作实验。该项目要求本地、区域和全球的人群与一个由艺术家、设计师和专家组成的团队一道,共同面对挑战并寻找答案——这个城市(及其不同的公众)应当如何应对景观变化过程。

设计的关键是创建一个开放的空间,促进来自城市各个角落各类人群(物种)之间的理解和交流。这个基地位于人群与河流交汇的关键节点。设计需要通过了解场地人群(物种)的构成和它们之间的关系提供适合不同类型公众的不同项目,使场地成为一个平等共荣的空间。

Through the understanding of these diverse publics, the lab is defined as collaborative experiment in landscape design. The project asks local, regional, and global communities to join together with a team of artists, designers, and experts to face a key challenge to the contemporary city and provide an answer. How does the city (and its diverse publics) respond to the processes of landscape change?

The key to the design is to create an open and welcoming space, and promote understanding of the diverse and multiple publics that comprise any urban area. This site is located at a critical junction between rivers and people. Design needs to develop an understanding of whom those publics are and how they interact to provide the types of diverse while creating spaces that do not exclusively privilege one group over another.

概念生成
Towards Port Lab

LANDSCAPE REGENERATION | THE REVITALIZATION OF THE WATERFRONT PORTAL

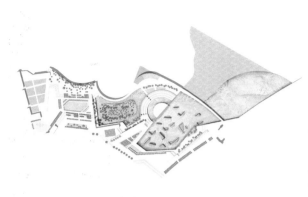

2037 年平面
Masterplan 2037

未来更频繁严重的降雨使海平面升高，淹没这个地点。更加漫长以及极端的潮汐现象开始在前军事基地的墙壁上沉淀沉积物。

More frequent and severe rain in the future begin to inundate the site with increasing sea level. Continuous and more extreme tidal action has begun to deposit sediment around the walls of the former fort.

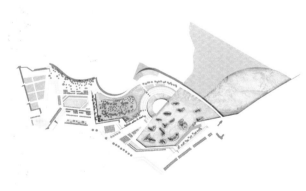

2067 年平面
Masterplan 2067

上升的海平面充分深入到这些地点的内部区域。潮汐带来了丰富的沉积物，产生了新的栖息地，使岛屿更为成型。

Rising sea level penetrates deeper into the site's inner area. Islands have been more fully formed and given rise to new habitats created by the rising of the tides and rich sediment environments.

 鵟 Buteo

 天鹅 Cygnet

 东方沼泽鹞 Eastern marsh-harrier

 鸳鸯 Mandarin duck

 鱼鹰 Osprey

 红隼 Kestrel

 铜锈环棱螺 Bellamya aeruginosa

 红螯螳臂蟹 Chiromantes haematocheir

 谭氏泥蟹 Ilyoplax deschampsi

 刺沙蚕 Neanthes japonica

 弹涂鱼 Periophthalmus cantonensis

 螳臂蟹 Chiromantes dehaani

 背蚓虫 Notomastus latericeus

 河蚬 Corbicula fluminea

保护鸟类
Birds(for Protection)

鱼类及两栖动物
Fishes & Amphibians

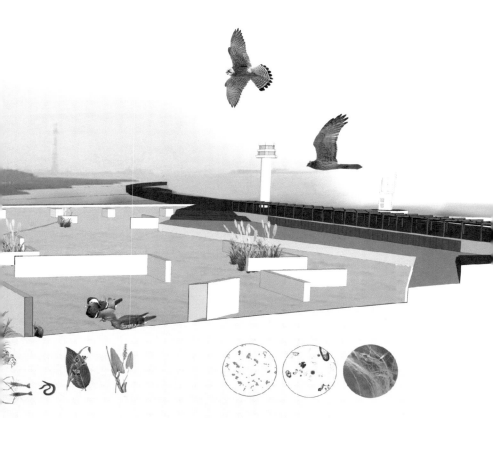

鸟类
Birds

鱼类及两栖动物
Fishes & Amphibians

浮游生物
Planktons

藻类
Algae

微生物
Microorganisms

良好的生态环境
Healthy Habitat Environment

引入
Introduce

食物链
Food Chain

地形
Topography

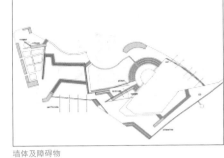

墙体及障碍物
Walls and Barrier

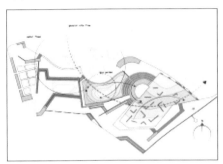

水流
Water Paths

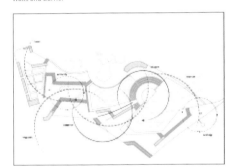

人流
Pedestrian Paths

剖面 A：穿过实验室总部、圆形露天剧场和湿地
Section A: Crossing the lab HQ, amphitheater and wetland

剖面 B：穿过湿地、河滨步道和碎石堆景观
Section B: Crossing wetland, riverwalk and a view through the scree

剖面 C：穿过雨水花园和上升步道
Section C: Crossing the rain garden and elevated walkways

LANDSCAPE REGENERATION | THE REVITALIZATION OF THE WATERFRONT PORTAL

重要的干预措施使该地区恢复了开放的状态,允许水流、野生动物和人们自由地深入到这个地点,构建河流的入口,重构人们之间的联系。该场地的配置开始受到熵的影响,并为城市创造了一个动态的新门户,它将持续吸引新团队,并产生新的变化。

The significant intervention has returned the site to an open form, allowing water, wildlife and people to penetrate freely deep into the site, which is able to restore access to the river-front and the connections between people. The configuration of the site immediately begins to be subject to the forces of entropy and create a dynamic new portal to the city that will continue to draw new teams and keep changing.

鱼菜共生
与潮为友
Aquaponic Aquaponding

组员 Members
林梦楠，王岱蕾，何沛文，Curtis Schaldach, Luisa Vogel, Antonia Besa

导师 Tutor
Jim Ayorekire

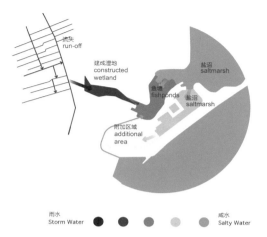

雨水　Storm Water　●　●　●　●　●　咸水　Salty Water

Aquaponic-Aquaponding 是一个有利于自然和建筑环境的循环。 一系列相互连接的盐水和雨水系统响应长江的日常潮汐流,有利于现场的水循环。生态、军事遗产、社区和地标的综合相互作用创造出一个景观,以纪念上海的海滨门户的过去、现在和未来。

Aquaponic-Aquaponding is a cycle that benefits both the natural and built environment. A series of connected salt water and storm water systems respond to the daily tidal flows of the Yangtze River for a beneficial circulation of water on site. The synthesis of ecology, military heritage, neighborhood, and landmarks interact to create a landscape that honors the past, present and future of Shanghai's waterfront portal.

暖水　Warm Water　●　●　●　冷水　Cold Water

LANDSCAPE REGENERATION | THE REVITALIZATION OF THE WATERFRONT PORTAL

雨水（建成湿地）
Storm Water (Constructed Wetland)

咸水（长堤）
Salt Water (Jetty)

咸水（池塘）
Salt Water (Pond)

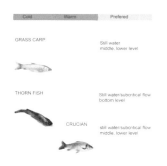

暖水（建成湿地）
Warm Water (Constructed Wetland)

冷水（长堤）
Cold Water (Jetty)

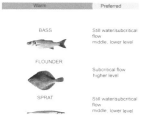

暖水（池塘）
Warm Water (Pond)

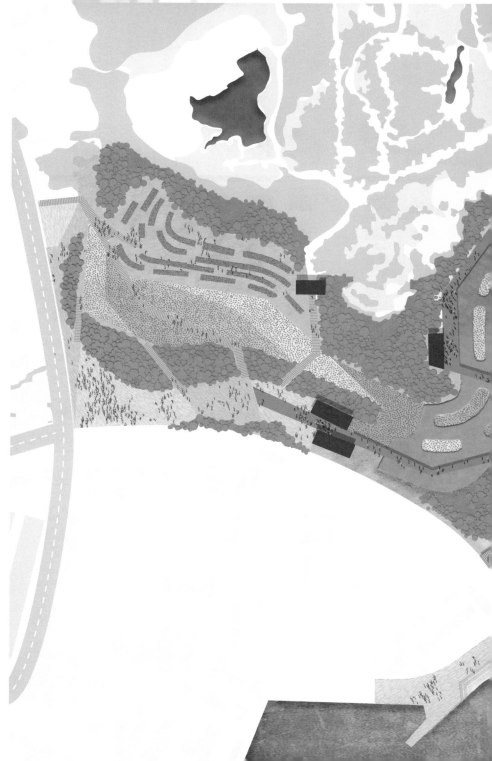

高潮
High Tide

总平面
Masterplan

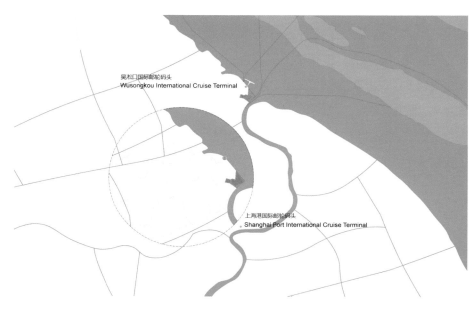

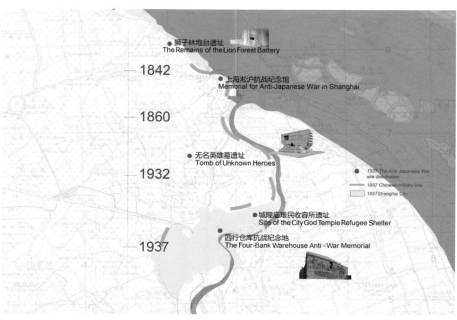

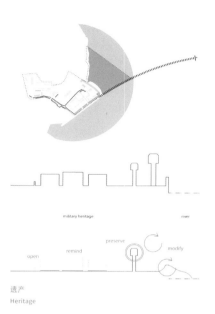

通过对历史资料的分析和解读,发现基地周围有丰富的历史脉络,包括狮子林炮台遗址、上海淞沪抗战纪念馆、无名英雄墓遗址等。因此,在场地设计中希望对场地中的保留建筑进行改造,结合场地周围的文脉进行更新,通过历史文脉的植入,使得场地重获新生。

By analyzing historic materials, the surroundings of the site are found to be of great historical value, including the Remains of the Lion Forest Battery, Memorial of Anti-Japanese War, the Tomb of Unknown Heroes. Therefore, it is expected to convert these remained buildings by implanting historical context, which may lead the site to the renewal.

遗产
Heritage

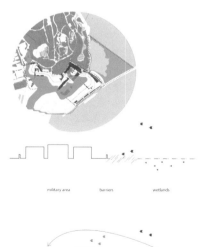

硬质化的基地边界使得海水和岸上的交流被阻隔,生物多样性减少,因此希望景观介入打破硬质驳岸的阻隔,将潮水引入基地,使得两侧连通,增加生物之间的交流,同时希望人工引入水生植物、鱼类等,增加场地的生物量,促进场地的发展。

The hardened frontier of the site cuts off the communication between land and sea, decreasing the biological diversity. Therefore, the intervention of landscape is expected to break the barrier, connect both sides and enhance communication among creatures. Meanwhile, it is also expected to introduce hydrophytes and fish artificially into the site. It may increase the sum of creatures and promote site development.

生态
Ecology

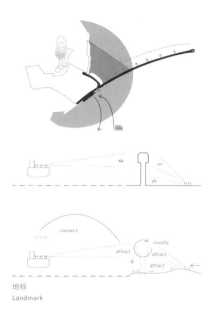

地标
Landmark

通过基地分析，我们发现基地具有很好的现状资源，例如长堤、高塔，但地形单一且缺乏标识性。因此，我们希望能够很好地利用场地资源，通过重新改造堤、塔以及重塑沿堤的地形来吸引岸上以及海上人群的目光。通过景观手段的介入，在增加场地的地标性同时，也提高了景观的辨识度。

By analyzing the site, we found the present resources hold excellent situation but the terrain is too simplex to have labeling function. Therefore, it is expected to attract people by rebuilding the bank, the tower, and reshaping the terrain along the bank. In the end, the intervention of landscape creates a both regional and recognizable icon.

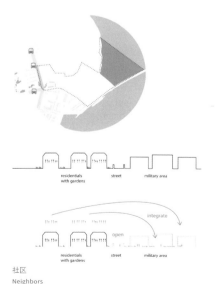

社区
Neighbors

设计前期我们调研并分析了周边的社区分布情况，发现周边社区的居民缺乏一个公共共享社区空间进行交流、休闲。因此，设计考虑在场地中建立一个共享的、充满活力的社区花园，以供社区居民交流合作，并将农场引入场地调动居民活动的积极性，从而打造一个极好的教育、休闲、交流的区域。

In preliminary design, through analyzing the distribution of neighborhood communities, we found the site is lack of a public and shared community space to communicate and relax. Accordingly, we designed a vibrant community park to provide a place for interacting and cooperating, and a farm to arouse enthusiasm of the residents. It will be an excellent place for education, leisure and communication.

LANDSCAPE REGENERATION | THE REVITALIZATION OF THE WATERFRONT PORTAL

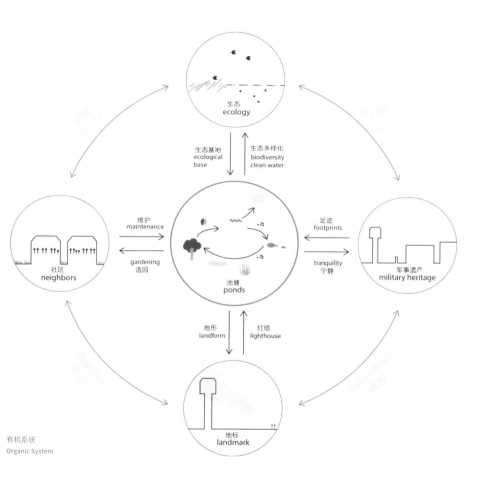

有机系统
Organic System

景观再生 | 滨水门户的活力复兴

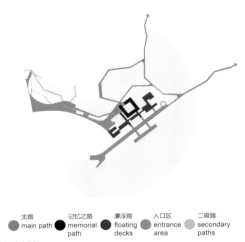

| 主路 main path | 记忆之路 memorial path | 漂浮路 floating decks | 入口区 entrance area | 二级路 secondary paths |

步行系统分析
Paths System Diagram

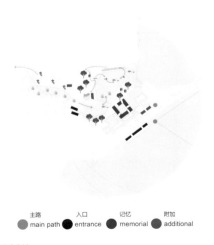

| 主路 main path | 入口 entrance | 记忆 memorial | 附加 additional |

活动分析
Program Diagram

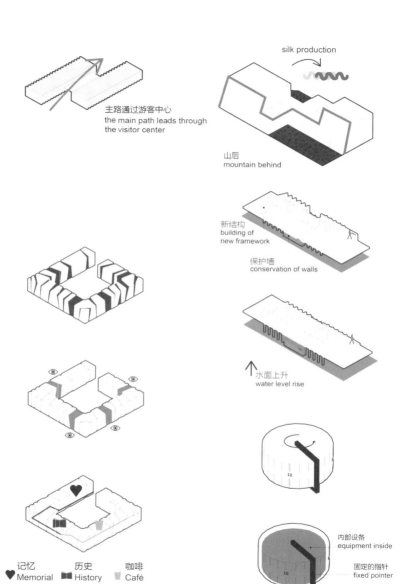

建筑分析
Architecture Diagram

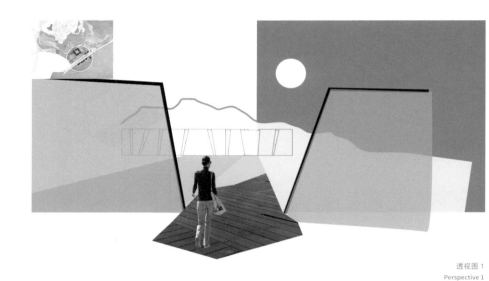

透视图 1
Perspective 1

透视图 2
Perspective 2

LANDSCAPE REGENERATION | THE REVITALIZATION OF THE WATERFRONT PORTAL

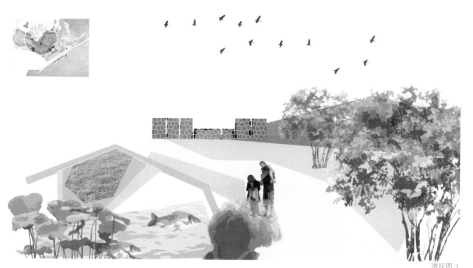

透视图 3
Perspective 3

透视图 4
Perspective 4

长江侧剖面
Section A Yangtze River

池塘系统剖面
Section B Pond System

LANDSCAPE REGENERATION | THE REVITALIZATION OF THE WATERFRONT PORTAL

我们希望通过对基地地形的重塑，增加场地地形的多样性，增加场地的标识性，同时增加生物的多样性。另外，通过池塘系统的设计，起伏的地形条件为生物创造绝佳的生长环境，满足生物生长的同时也能够增加人们游玩的趣味性，一举两得。

By reshaping the terrain, it is expected to enhance the terrain diversity and labeling function of the site, as well as the biodiversity. Meanwhile, the design of the pond system creates a perfect growing environment for creatures, which also can arouse people's interest in visiting.

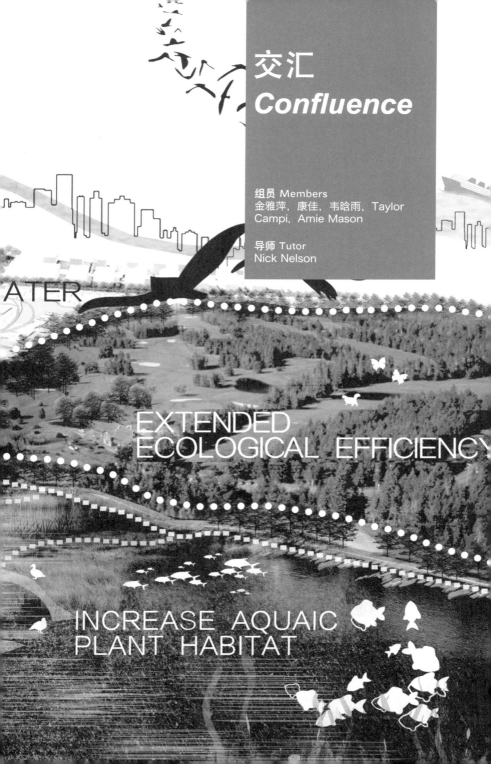

场地位于长江和黄浦江的交汇之处,虽有如此理想的地理位置,却因场地内高墙和军事用地的影响而交通不便。目前场地西侧用于居住和商业,南北均为连续的滨水绿地。从交通上看,北侧是开放的国际邮轮码头。另外,东北侧建立了防汛墙抵挡潮汐影响。经过分析,场地主要问题在于几乎没有公共空间,与周边的公园和社区皆不相连,植被单一,高墙和军事用地造成了视觉和物理上的隔断。

The site is located at the confluence of the Yangtze River and the Huangpu River. Owning such a prime location, it is currently disconnected due to tall walls and military facilities. There are residential and commercial areas in the west. Both the northern and southern part is the waterfront green land. In terms of transportation, there are wharfs along the river and the international cruise terminal in the north. The flood-control walls are in the northeast, preventing tides from the site. Therefore, the main problems of the site are lack of public space, disconnected from surrounding park and community, monotonous vegetation and inaccessible both visually and physically.

现状土地利用
Current Land Use

现状交通
Current Transportation

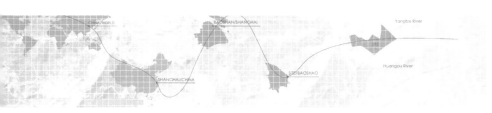

LANDSCAPE REGENERATION | THE REVITALIZATION OF THE WATERFRONT PORTAL

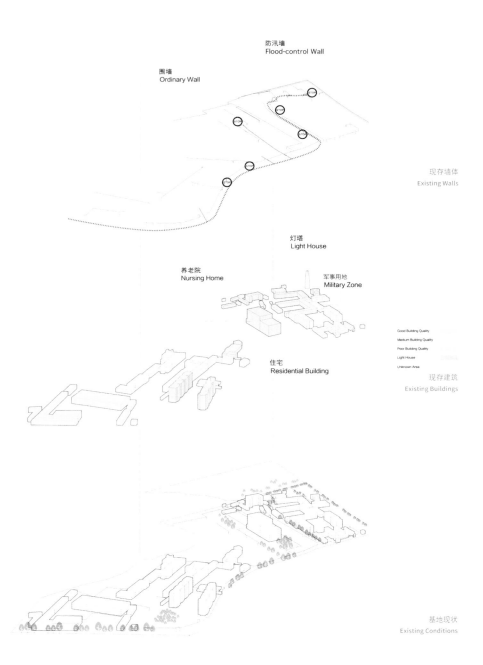

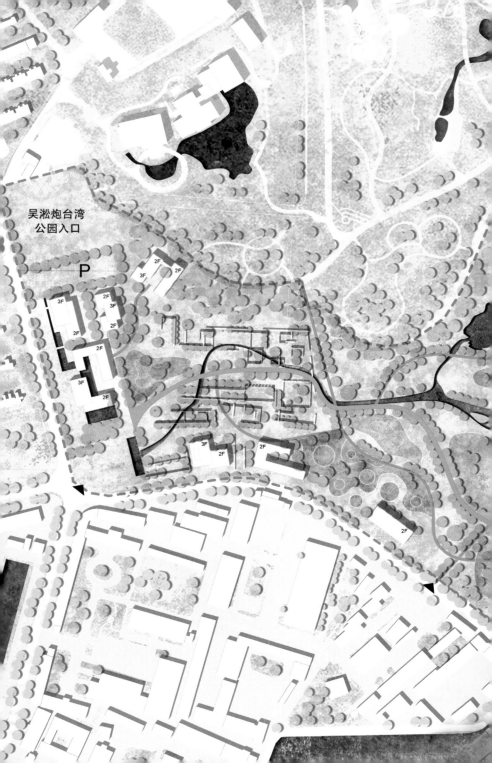

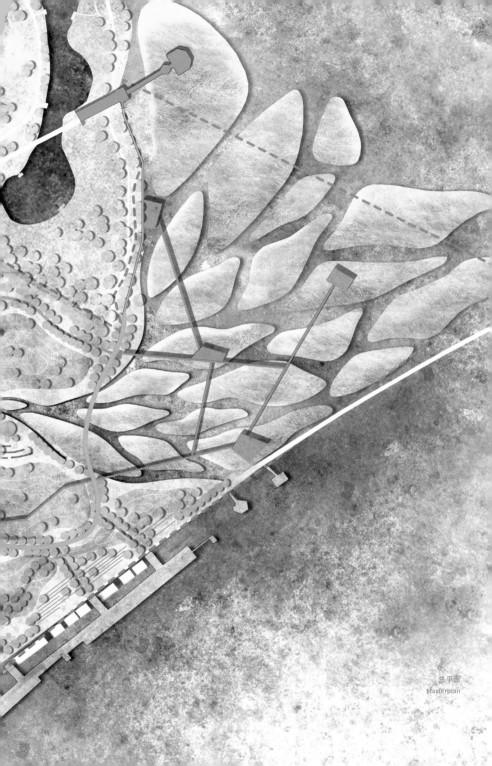
总平面
Masterplan

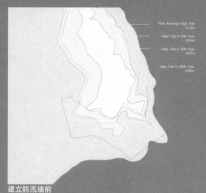

建立防汛墙前
Without the Flood-control Wall

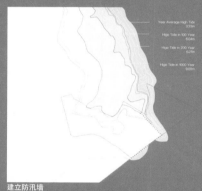

建立防汛墙
Built the Flood-control Wall

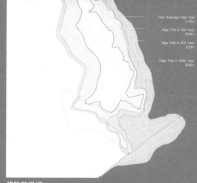

移除防汛墙
Remove the Flood-control Wall

方案使用交汇的概念，在上海最重要的两条河流交汇处，将人们重新连接起来。该地块目前可达性差，而未来将作为公共活动、生态恢复以及与自然联系的空间而重新焕发生机。城市地表雨水收集系统引导游客在场地中游览，城市的空间成为潮汐湿地。通过对现有建筑的局部保留及功能置换，营造更多的公共艺术空间和亲水空间。通过将经植被净化的雨水引入湿地，游客可以亲近生态系统，在草地露台以及漂浮的咖啡厅休憩。交汇不仅是两条河流交汇的地方，也是城市生活和生态的交汇点，为世界各地的人们创造了独特的财富。

The proposed plan uses the concept of Confluence to reconnect people with their waterfront at the point where two of the most important rivers meet in Shanghai. The site, though inaccessible and disconnected for now, will be revitalized as a space for public activity, ecological restoration, and connection to nature. As daylighting storm water leads visitors through the site, urban spaces become regenerated tidal wetlands. Repurposed building remnants, public art, engagement with water, and other experiential features evoke images of confluence throughout the site, while vegetation cleanses the storm water stream that leads into the wetlands. Visitors may enjoy an intimate and low-impact experience of the ecosystem or connect with the waterfront on grassy terraces or a floating cafe. "CONFLUENCE" is not only the point where two rivers meet—but also where urban life and ecology flow together to create a distinct community with assets for people from around the globe.

LANDSCAPE REGENERATION | THE REVITALIZATION OF THE WATERFRONT PORTAL

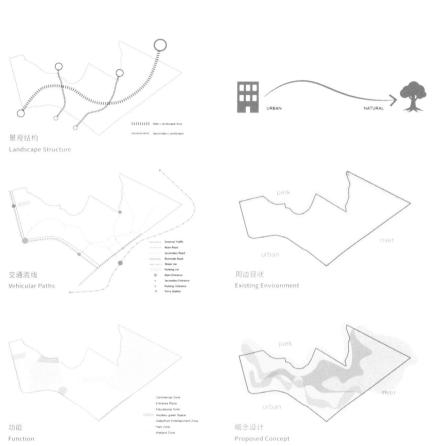

概念"交汇"有效地将基地周边的城市、公园和水体联系起来。西南面的城市商业和居住区域将渗透进北面和东面更为绿色的公园、湿地和水岸地带。随着人向东漫步体验，城市氛围逐渐减弱而景观愈发接近自然。因此，"交汇"的概念让人与水体重新建立联系，并通过创造开放空间以及健康的生态系统将基地重新激活。

The concept Confluence effectively blends together the three sides of the site—urban, park, and waterfront. The urban commercial residential areas in the southwest will bleed into greener conditions in the north and the east, where parks, wetlands and waterfront are. As visitors move eastward through the site, it gradually becomes less urban and more natural. The new design uses the concept of Confluence to reconnect people with waterfront and revitalize the area by creating active public spaces and healthy restored ecosystems.

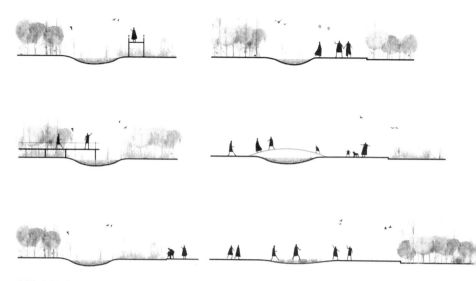

路径与水之间的联系
Relationship between paths and water

场地剖面 1
Section 1

场地剖面 2
Section 2

人与自然的联系从一进入场地就开始建立。点状绿地将自然植入沿街的商业空间,贯穿基地的地面排水系统将暴雨径流引入地下,因而基地的水特征得以强化。

树木、藤本、花卉以及草地蔓延渗透至功能重置后的建筑残骸以及玻璃盒子。不同层次的木栈道使得人与自然产生多样的联系——有的在树荫之下,贴近于地表;有的穿梭于树林,甚至行走于树木之上。路径将引导人们沿着溪流,或跨越溪流(桥),或穿过溪流(汀步),到达森林和绿色开放空间以及水岸的湿地。木栈道系统架空于湿地,因而在近距离体验生态的同时,能够减少对栖息地的人为影响。

Contact with nature begins at the site entrance, where small gardens bring nature into the commercial space, and the underground storm water is to begin its journey through the site.

Trees, vines, flowers, and grasses grow around and into and over the repurposed building remnants and their glass additions. Multiple boardwalk levels allow visitors to interact with nature in different ways—some at ground level, covered by trees, and some in and above the tree canopy. The path takes visitors alongside the stream, over the stream (as bridge), or through the stream(as stepping stones), and leads to areas of dense forest and open green spaces toward the wetlands and waterfront. A system of boardwalks runs over the wetlands (multi-level in some places) so that people can experience this ecosystem up-close with a low impact on the habitat.

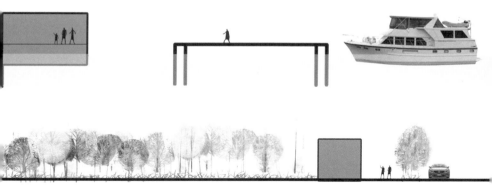

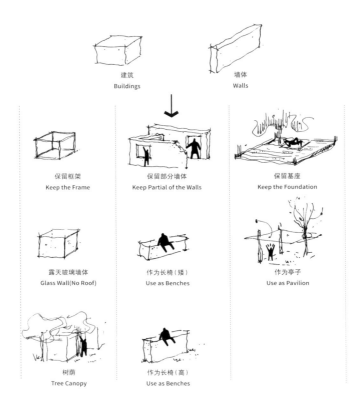

现状建筑与墙体处理策略
Strategies to use the Existing Buildings and Walls

保留框架，植入玻璃墙，环绕以森林，重新利用为展示空间、自然教育场所（植物及花卉）；保留部分墙体，使得新与旧（玻璃与残骸）之间产生对比；保留可再利用为座椅和休息亭的构件。

Keep the frame and implant glass walls surrounding with trees in order to reuse the site as an exhibition place and educational venues for nature (plants & flowers). Then, keep some walls to make a contrast between the new and the old (glass and remnants). Finally, others can be used as benches or pavilions.

水体蜿蜒穿过建筑内外,引导人们进行探索:人们可以视线透过墙体(窗洞)、行走穿过墙体(门洞)、踏上台阶欣赏不同视角的景观,感受各种材质(重的砖块、轻的玻璃、生长的植物、流动的水体)的质感。

As the water goes through the buildings, it will lead people to explore this area— people seeing through the walls (windows), walking through the walls (door), going up stairs to have multiple views of surroundings, to experience different textures (heavy brick, light glass, growing plants, flowing water) in this area.

人、墙体与水之间的联系
Relationship between people, walls and water

场地剖面 3
Section 3

场地剖面 4
Section 4

LANDSCAPE REGENERATION | THE REVITALIZATION OF THE WATERFRONT PORTAL 65

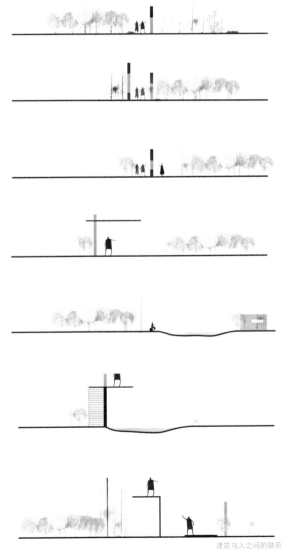

建筑与人之间的联系
Relationship Between People and Walls

LANDSCAPE REGENERATION | THE REVITALIZATION OF THE WATERFRONT PORTAL

拆除部分防汛墙，连接暴雨径流和江水，使湿地系统与北侧森林公园建立联系。选择具有净化功能的水生植被，并为野生动物提供栖息地。下游植被应更加耐盐，低耐性植物应放置在上游生长，湿地物种则在东部生长。湿地可以作为栖息地庇护生活在沿海湿地的珍稀鸟类。

Demolish flood-control walls, connect stormwater with brackish riverwater, connect the wetland system to the park in the north. Intentional stream vegetation for cleaning dirty water, provide habitat for wildlife. Downstream plants should be chosen from those which are more salt-tolerant. Less tolerant plants grow upstream while the brackish wetland species thrive down in the east. These wetlands may serve as habitat for some rare birds that live in the coastal wetland.

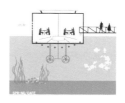

春 / 咖啡
Spring/Cafe

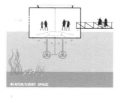

冬 / 活动
Winter/Event

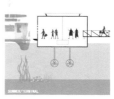

夏 / 航运
Summer/Terminal

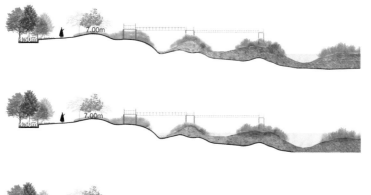

平均低潮水位 0.93 米
0.93m Average Low Tide Level

平均高潮水位 3.33 米
3.33m Average High Tide Level

两百年一遇洪水位 6.21 米
6.21m 200-year Flood Level

Alo'ha Isla

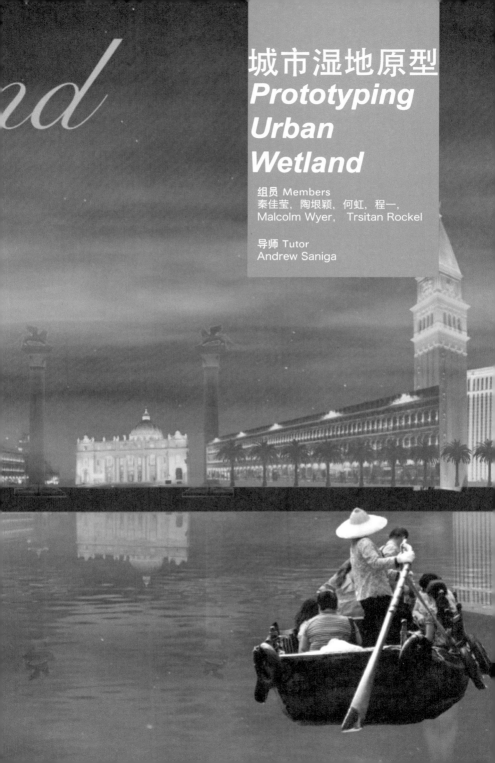

城市湿地原型
Prototyping Urban Wetland

组员 Members
秦佳莹，陶垠颖，何虹，程一，
Malcolm Wyer, Trsitan Rockel

导师 Tutor
Andrew Saniga

LANDSCAPE REGENERATION | THE REVITALIZATION OF THE WATERFRONT PORTAL

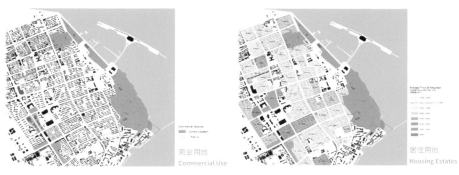

商业用地
Commercial Use

居住用地
Housing Estates

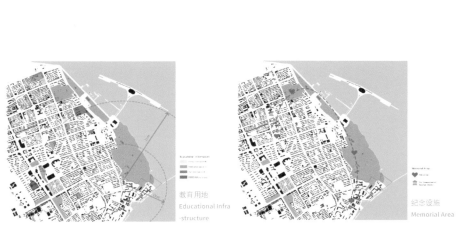

教育用地
Educational Infra-structure

纪念设施
Memorial Area

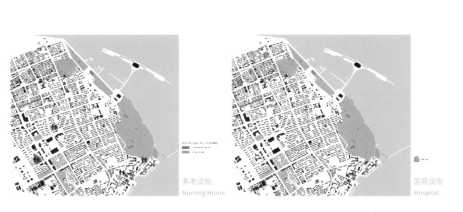

养老设施
Nursing Home

医院设施
Hospital

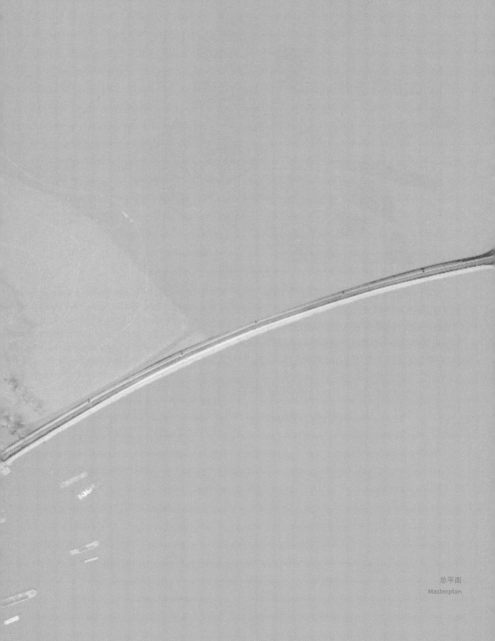

总平面
Masterplan

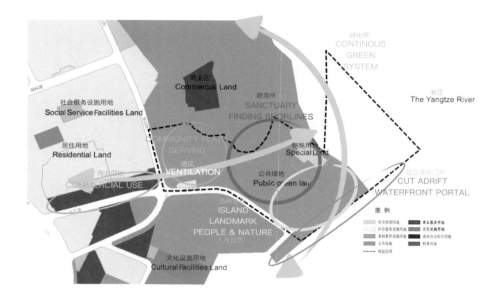

几个世纪以来的城市发展侵占了湿地,如今应把河水编织进上海的城市肌理。在高密度城市的环境下,河口盐沼不仅保护着重要的生态生物多样性,而且给人类带来景观和游憩。曾经不受欢迎和重视的沼泽景观,现在反而为宝贵的滨水区发展提供了新的机会。我们认为,河流恢复不应局限为湿地恢复,经过设计的湿地可以淹没新的陆地,"洪水"是许多用来建立新的滨水区的设计方式之一。

After centuries of urban development encroaching upon China's wetlands, the time has come to weave water back into the urban fabric of Shanghai. Estuary salt marshes not only support critical ecological biodiversity but also support scenic and recreational opportunities for people stressed by high-density urban sprawl. In turn, landscapes once viewed as undesirable swamps now offer a new opportunity for supporting valuable waterfront development. This project argues that river restoration should not be seen as the return of wetlands to their pre-urban form. Instead, designed wetlands can flood new terrestrial zones, with "flooding" functioning as one of many design operations utilized to establish new waterfront neighborhoods.

LANDSCAPE REGENERATION | THE REVITALIZATION OF THE WATERFRONT PORTAL 75

景观再生 | 滨水门户的活力复兴

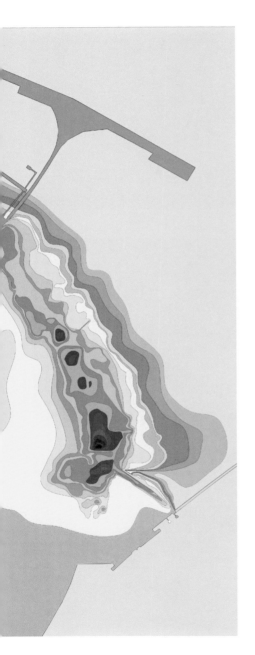

潮汐分析
Analysis

LANDSCAPE REGENERATION | THE REVITALIZATION OF THE WATERFRONT PORTAL

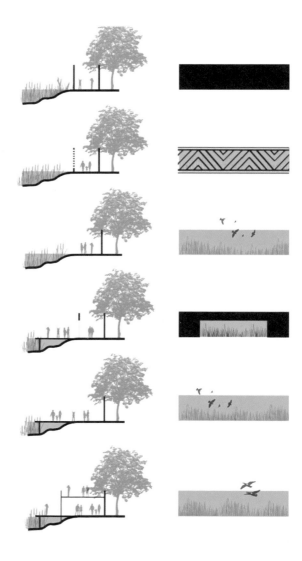

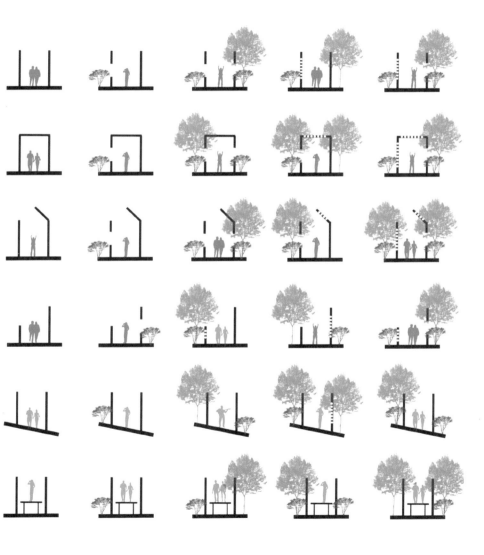

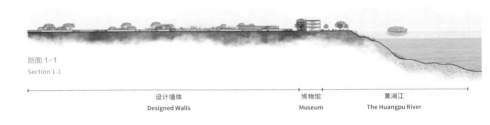

剖面 1-1
Section 1-1

| 设计墙体 | 博物馆 | 黄浦江 |
| Designed Walls | Museum | The Huangpu River |

剖面 2-2
Section 2-2

| 盐沼 | 生态展示中心 | 岛 | 黄浦江 |
| Salt Marsh | Ecology Exhibition Center | Island | The Huangpu River |

LANDSCAPE REGENERATION | THE REVITALIZATION OF THE WATERFRONT PORTAL

剖面 A-A
Section A-A

剖面 B-B
Section B-B

剖面 C-C
Section C-C

在黄浦江和长江的汇合处,有一个以人工填土为基础的前军事和工业基地。建议在该地的北部陆地边界建立一个经过设计的潮汐河口。阶梯式的地形创造了直线围合的水面图形,并随着潮汐的波动,转变成新的形式。通过展现生态生长的过程,盐沼促使人类与自然接触。这种扩张湿地,创造新的滨水区的方式,将废弃的工业景观变成"Alo' ha*岛"。因此,这个项目提供了一个景观改造策略,可将原型沿黄浦江地区一直拓展到上海中部,创造一个绿色的、不断扩大的滨河走廊,促进文化、生态、商业交换。

At the confluence of the Huangpu River and Yangtze River, it lies a former military and industrial site built on artificial fill. This project proposes the establishment of a designed tidal estuary on the northern terrestrial boundary of the site. Terraced landforms create rectilinear geometries of water that transform into new shapes with tidal fluctuation. By making ecological processes visible, this salt marsh engages human populations with nature. This expanded wetland creates new waterfront districts, transforming a derelict industrial landscape into "Alo'ha Island". This project offers a landscape transformation strategy that could be prototyped along the Huangpu River all the way to central Shanghai, creating a continuous green corridor of expanded riverfront, encouraging cultural, ecological, and commercial exchange.

*Alo' ha: 夏威夷语,本有希望、爱、和平以及幸福等意思。19 世纪中期被纳入英文词汇成为问候语,与"你好"的意思相近。

LANDSCAPE REGENERATION | THE REVITALIZATION OF THE WATERFRONT PORTAL

惊奇 韵律 释放
Riddle
Rhythm
Release

组员 Members
沈萱，郭屹，王依桐，Ayano Healy, Mendez Garcia, Lucia Tilling

导师 Tutor
董楠楠

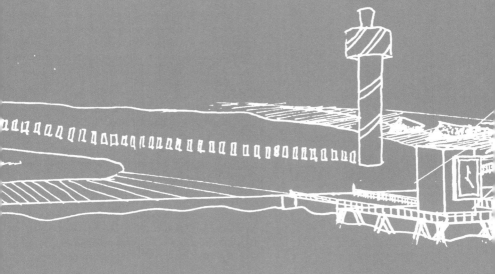

上海宝山吴淞口滨水门户位于黄浦江河口,是黄浦江沿岸功能网络中的重要一环。在黄浦江滨江带更新的进程中,加强沿江各功能枢纽的相互联系,进而促进城市与滨水地段的联系,这一愿景有着极大的发展潜力。

吴淞口滨水门户处于宝山区各大功能版块之间的地段。目前,场地内部主要为军事用地,这极大地割裂了场地与城市的关系。因此,场地亟待更新以促进宝山区各功能版块的协同发展,从而改善地区的连通性,增强地区活力以及塑造地区特性。

上海总体规划中提出了让城市更具活力、吸引力,并能可持续发展的目标。因此,规划设计中希望置入新颖的活动来促进人们的相互交流,从而为场地注入生机与活力。

The Baoshan Waterfront Portal is part of a larger city network of themed hubs that line the Huangpu River. As the city continues to activate sites along the Huangpu River, there is a potential to improve connectivity between the different hubs. This creates an opportunity to amplify Shanghai's connection with the waterfront.

The Baoshan Waterfront Portal is located at the intersection of different functions within the district. Currently, the site is occupied by the military, making the site absent and deteriorated. However, the site has a potential to redevelop in synergy with the surrounding functions, thereby improving connectivity, urban life and the identity of Baoshan District.

The Shanghai City Masterplan aims to make the city more dynamic, attractive and sustainable. The proposed strategy for the Baoshan Waterfront Portal promotes people's interactions by introducing new activities, creating a series of experiences that will make the site more attractive and vibrant.

LANDSCAPE REGENERATION | THE REVITALIZATION OF THE WATERFRONT PORTAL

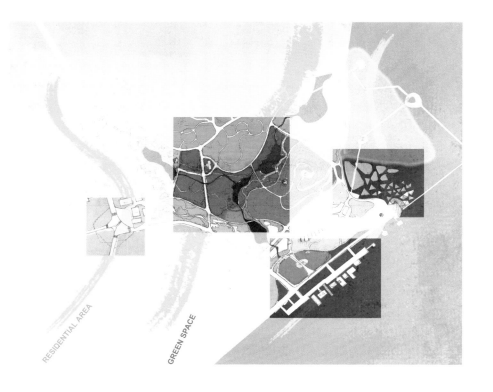

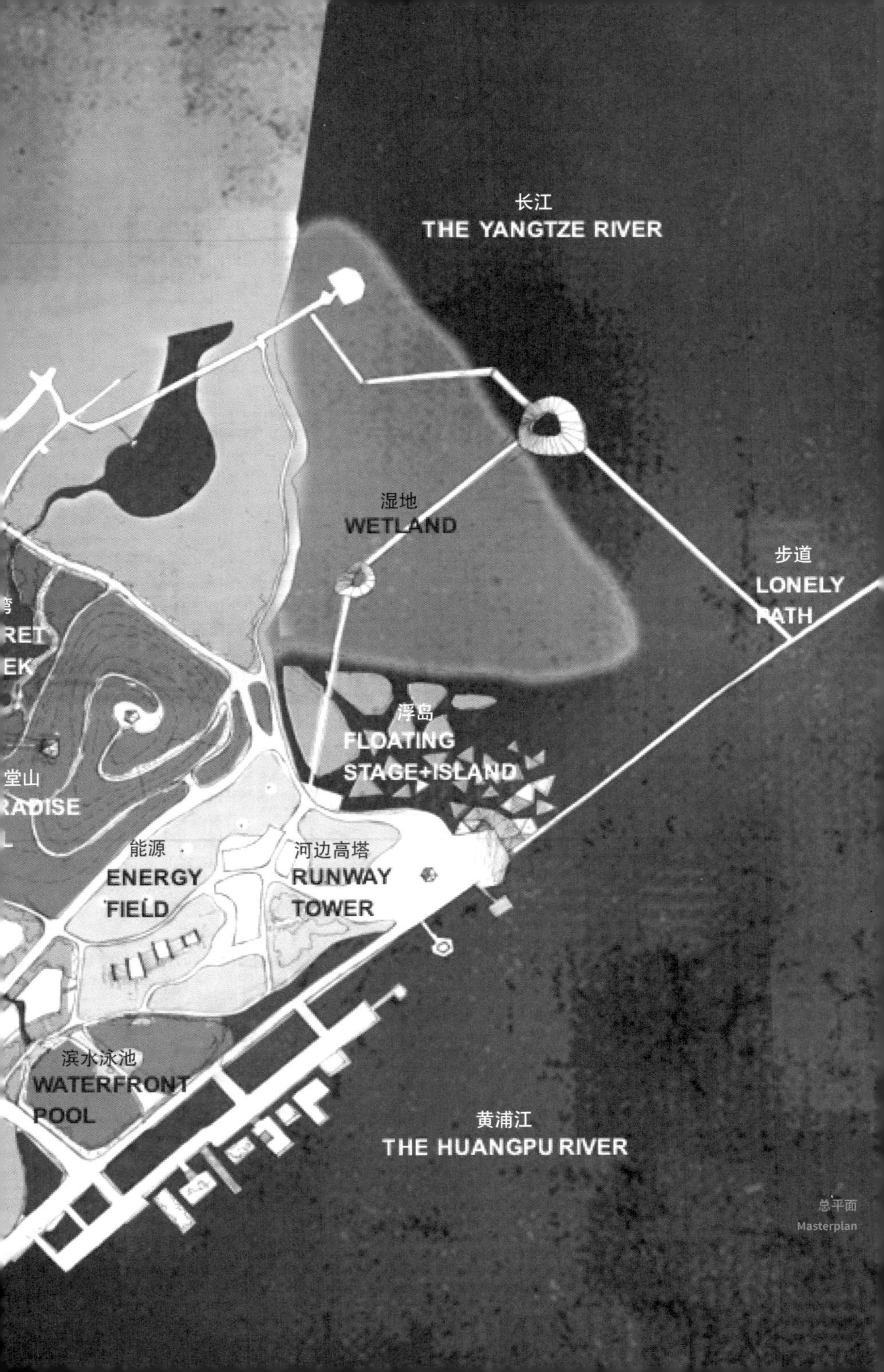

长江
THE YANGTZE RIVER

湿地
WETLAND

步道
LONELY PATH

浮岛
FLOATING STAGE+ISLAND

能源
ENERGY FIELD

河边高塔
RUNWAY TOWER

滨水泳池
WATERFRONT POOL

黄浦江
THE HUANGPU RIVER

总平面
Masterplan

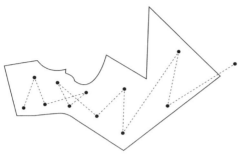

设计概念为"惊奇 韵律 释放"。"惊奇"意为在场地中穿行所获的惊奇的场景感受;"韵律"意为在场地中游历所经历的有节奏的空间变化;"释放"意为清理场地现有废旧元素,建立新的舒适自由的生活方式以及改善自然环境。

设计意在复兴宝山区滨水门户,加强人与人之间的相互交流。在这里,游客与本地居民能通过一系列的活动交融在一起,并能以一个独特的极具活力的视角认识上海这座城市。

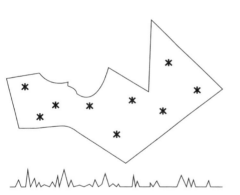

The concept Riddle, Rhythm and Release gives life to the actions within our strategy. Riddle speaks about the surprises that come along with discovering a new place. Rhythm refers to the emotional peaks, valleys and climax that people experience as they visit the site. Release is the action of clearing the site of existing construction to make new connections, bring life in and improve the natural environment.

The Waterfront Portal aims to revitalize the Baoshan district and create a space that emphasizes interaction between people. Locals, as well as tourists, will be able to enjoy the array of activities available and discover Shanghai from a dynamic perspective.

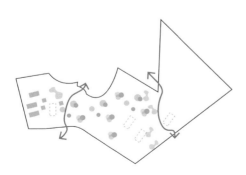

LANDSCAPE REGENERATION | THE REVITALIZATION OF THE WATERFRONT PORTAL 93

孤独小径 1
Lonely Path 1

孤独小径 2
Lonely Path 2

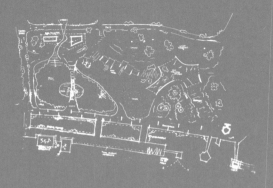

滨水泳池是场地的一个标志性景点。通过一套水池过滤净化生态系统，水体最终汇入滨水泳池。丰富的水上活动将更好地让城市生活与水上生活相联结。

场地滨水区包含一系列漂浮的模块化小岛。人们可根据自己的意愿变化组合这些模块。这些模块小岛的类型包含：水上栈道、休憩平台、活动广场和景观建筑。

A key landmark feature of the Waterfront Portal will be a waterfront pool. This will be created through natural filtration via a series of ponds, leading to the pool. It will offer an experience unprecedented in Shanghai, combining city life with water recreation.

The site includes a series of floating platforms that will be arranged next to the wetland. They are modular, which allows for flexible uses and adaptability. Possible uses include water paths, playgrounds, and landscape designs for habitat creation.

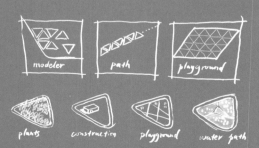

设计赋予了场地空间过渡的特性，从高密的现代城市空间转变到静谧的丛林，最后到开放、多元的滨水活动区。有节奏的变化契合了"韵律"这一概念。它引导人们从喧闹的城市中融入到宁静的自然空间，并让整个空间体验极富节奏感。

The project creates a smooth transition from an urban, densely developed area to a more open, multifunctional zone dominated by green space. This embodies the Rhythm mentioned in the concept. It leads people from loud urban space to more tranquil natural space, providing consistent points of interest throughout the site.

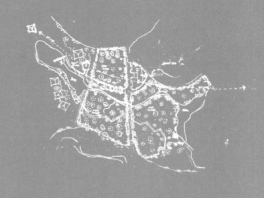

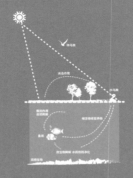 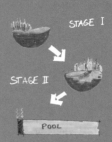

滨水区设计了一系列生态栖息地，分布在多样的植被群落中，从草坪到灌木，从山丘到丛林。它们相互作用建立起一个多样的生态群落，并为鱼类、鸟类等物种提供栖息地。

A series of habitats are created within the Waterfront Portal. These are established through a vegetation pattern that runs through the site and moves from meadows to small shrubs, larger pioneering trees, a climax woodland and loops back to the meadow. This ensures a broad spectrum of habitats for a large number of species.

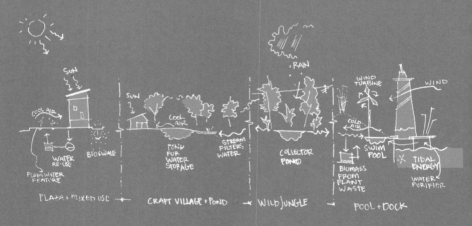

我们设计了一套自然能源生态系统，包含太阳能、风能、潮汐能和生物质能。通过在场地不同地段实施不同的生态技术及策略来收集这些生态能源。

The Baoshan Waterfront Portal is designed to make use of natural energy sources, such as solar, wind, tidal and biomass. The energy from these sources is captured by incorporating technologies into the different elements within the park.

剖面 1-1
Section 1-1

剖面 2-2
Section 2-2

剖面 3-3
Section 3-3

剖面 4-4
Section 4-4

剖面 5-5
Section 5-5

吴淞生态结
Wusong Eco-Nexus

组员 Members
张雯珺，林诗琪，徐玮蓬，Jung Jea Hyun, Brandon Crawford, Philipp Steinbacher

导师 Tutor
戴代新

浦江两岸的空间特性极为不同，宝山滨江区域以其邮轮产业、生态与旅游业为特色；黄浦江中段已实现贯通，宝山滨江所在的黄浦江北段的贯通势在必行。

沿浦江两岸现有 39 个码头，其中 2 处邮轮码头，4 处游船码头，3 处游船码头兼轮渡码头，其余皆为轮渡码头。

码头
Terminals

LANDSCAPE REGENERATION | THE REVITALIZATION OF THE WATERFRONT PORTAL

The waterfront spaces on different sides of the Huangpu River represent diverse characteristics. The Baoshan Waterfront features ferry industry, ecology and tourism. Since the middle part of the Huangpu River has realized linking, the northern part, where the Baoshan Waterfront lies, will also soon be accomplished.

Now it exists 39 wharves along the Huangpu River, including 2 cruise terminals, 4 marinas, 3 other marinas which also work as ferry terminals and the rest 30 ferry terminals.

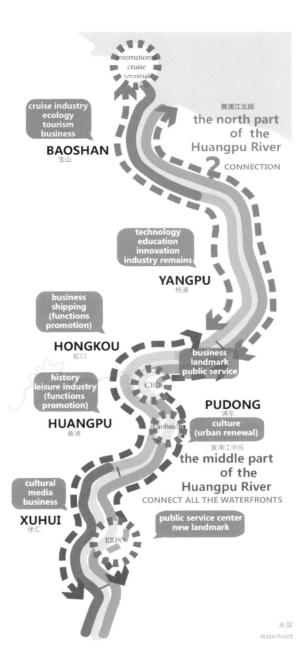

功能结构
Functional Structure

交通
Transportation

整体概念
Overall Concept

设计选定区域位于多种场地要求的交汇处。在区域整体规划中，可以明显看出该设计片区位于各类中心（区域中心、母港核心区、南部服务区）的边界区，因此在设计中对于该片区的功能定位以及如何整合丰富的周边场地要素成为设计重点。

从场地现状的交通情况可以看出以下问题：①场地的可达性较差；②场地与周边的公共交通联系较为松散；③设计选定区域与周围核心区存在一定阻隔，交通不便；④滨水景观不连续。另外，从目前公共汽车的路线及站点的设置，可以得出目前该片区核心区域不明显，功能分区尚未成形等特点，因此在设计过程中应当考虑该片区的未来发展趋势，以明确设计区域的功能定位。

设计区域纳入周围场地要素（滨水景观、社区、港口），并进行功能延伸（自然生态、社区活动、副港口及商业），在设计区域把这些功能积极地整合以求形成丰富的、具有活力的、非中心类活动片区，给居民提供娱乐场地，给市民提供生态游览体验区，给游客提供进入城市新的交通途径及了解上海市民生活的窗口。

The selected area is located at the influence of various required sites. In the master plan of this region, it is clear to see that the site is on the border of several kinds of functional centers. Therefore, how to define the function of the site and to integrate the rich resources around it becomes the most important issue in our design.

By observing the current situation of public transportation, there are four main problems, including poor accessibility, loose connection with neighboring public transportation, restriction from the central functional region and discontinuous waterfront landscape. Meanwhile, the setting of bus stations and routes also represents the unclear function of the site. Therefore, the trend of future developing should be taken into consideration for a better design.

The design integrates the neighboring site elements like waterfront landscape, community and harbor, and expands them to nature, public events and commercial uses. The selected area is designed to form a positive integration of several functions, so that it can be revitalized as a varied and lively non-central activity area. It will not only be a recreation place, but also an eco-tourism determination and a new path leading to the city and the life of citizens.

总平面
Masterplan

联系绿地
Connect the Green

联系人群
Connect the People

整合自然与人文景观
Integrate Natural and Social Landscapes

生态策略
Ecological Strategies

LANDSCAPE REGENERATION | THE REVITALIZATION OF THE WATERFRONT PORTAL

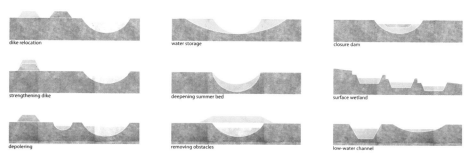

水的脉搏 / 洪水 / 干旱
Waterpulse/Flooding/Drought

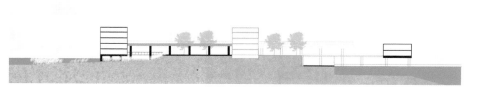

剖面二
Section 2

吴淞生态结整合了创新的景观设计，旨在通过旧建筑改造与潮汐湿地的恢复使宝山地区重现活力。长江与黄浦江在此交汇，上海与东海在此相遇，本次改造通过强化这种关系彰显基地的特色，连接了该地区的重要生态走廊。由此，该项目激发了多种生态用途，包括娱乐活动、自然教育、研究设计、社区参与和休闲体验，反映了上海的动态特质。吴淞的生态结将推进上海城市环境、经济和社会目标的实现，同时也为上海复兴了一个面向全球的城市门户。

The Wusong Eco-Nexus integrates innovative landscape design, repurposed buildings, and restored tidal wetlands to rejuvenate Baoshan District. This project leverages the site's unique location by emphasizing the intersection, or nexus, of the Yangtze River Estuary, the Huangpu River, Urban Shanghai, and the East China Sea. As such, Wusong Eco-Nexus strategically and seamlessly links the region's essential corridors. The project thereby reflects Shanghai's dynamic nature by integrating and stimulating diversity of ecologically oriented uses, including recreational activities, educational opportunities, research and design, community engagement, and leisure. The Wusong Eco-Nexus approach therefore assists Shanghai in advancing the city's environmental, economic, and social goals while providing a revitalized entrance to the world.

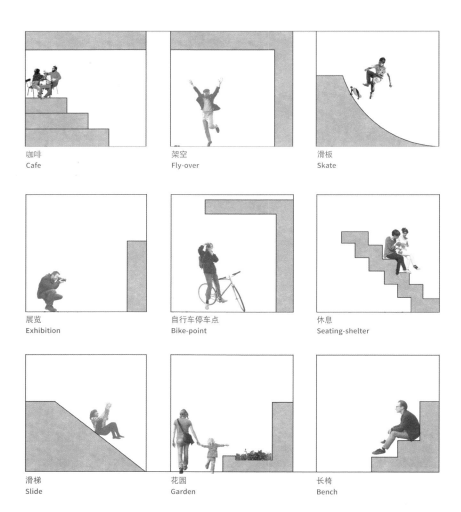

复合型运动场地
Multi-sport Arena

露天音乐会
Concert

城市花园
City-gardening

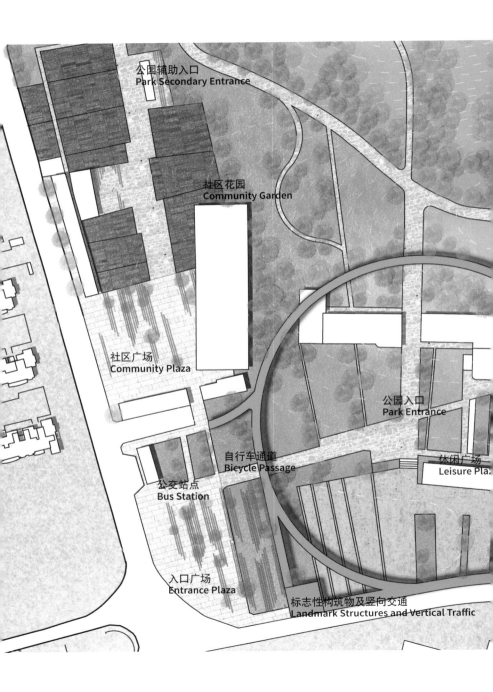

LANDSCAPE REGENERATION | THE REVITALIZATION OF THE WATERFRONT PORTAL

休闲港湾
Leisure Harbor

活动中心
Activity Center

湿地公园
Wetland Park

社区花园
Community Garden

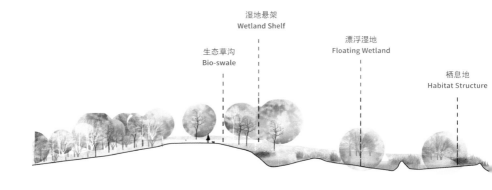

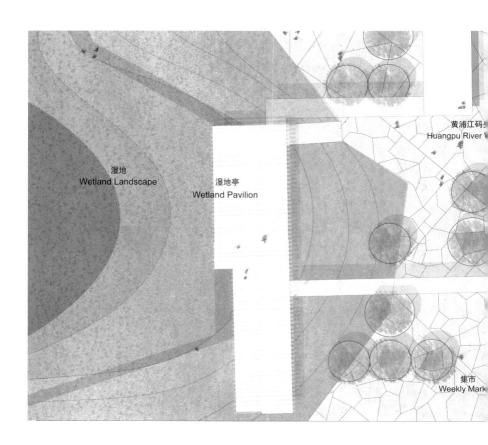

LANDSCAPE REGENERATION | THE REVITALIZATION OF THE WATERFRONT PORTAL

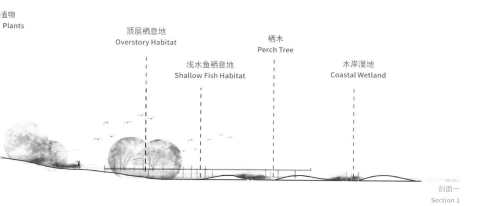

剖面一
Section 1

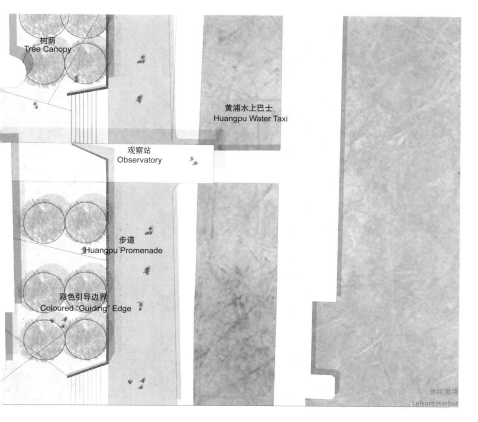

休闲港湾
Leisure Harbor

后记
Postscript

水边的感悟和交流
Inspiration and Communication Beside the Water

胡玎

作为一次国际夏令营,其目的在于"以点带面"地推动同济大学与国内外院校的交流。设计课题是夏令营这个"点"中的关键部分。以"滨水景观再生"作为设计的主题,以黄浦江口这片区域作为设计的对象,折射出组织者的匠心。在一年后的同济大学风景园林三年级本科生课程设计中,也引入此课题开展教学实践,我作为该课程的指导老师之一,对这一课题又有了一些新的认识和想法。下文以"水边的感悟和交流"为名略谈几点。

城市聚焦

夏令营中的成员从世界和全国各地聚集到上海,要结合课题让他们立足上海开展交流,认识上海、思考上海、建言上海。位于长江口是上海在全国的地理特征,奠定了其在长江经济带的龙头地位和责任。黄浦江是上海的母亲河,在方方面面孕育了这座城市。在两水交汇处,有着城市发展的诸多典型问题。尤其对于那些初到中国和上海的外国学生,通过这一课题从宏观到中微观认识和思考上海是很好的切入点。

在本科课程中,风景园林落实边做边学的专业特点,结合规划设计引导本科生在解决实际问题中入门,走上专业之路。这一课题便于学生脚踏实地,在上海中心城区的典型性区域思考上海发展中的真实问题。

专业聚焦

同济 CAUP 国际设计夏令营是由建筑系、城市规划系、景观学系轮流举办的学术交流活动。作为景观学系组织的一次夏令营,此课题引入了城市中天人合一的视角。这片区域中既有废弃工厂、码头等,需要进行旧城改造和城市功能更新;又有两水交汇和潮汐现象的独特水文,需要保护自然水体、湿地和野生的鸟类。在这一典型区域,从专业上回答如何实现上海自然和人文的可持续发展,提出各个团队的建议和方案,很有意义。

在本科课程中，同济大学风景园林专业三年级的本科生正处于专业学习的转型阶段。一、二年级时积累全面和多元的规划设计专业基础，从三年级开始更清晰地进入到风景园林规划设计的培养中，以公园、绿地系统、风景名胜区规划设计等为载体进行学习。此课题很符合专业学习转型阶段需要兼顾生态与建筑、自然环境与人工环境的特点。

时间聚焦

夏令营中，国内外学生是在一个特定的时间点上开展学术的交流。2017年是上海市贯通黄浦江滨江岸线的冲刺时段。上海是中国近代史上民族工业创办和发展的重要城市。黄浦江沿岸码头、仓库、工厂林立，见证了这一历史过程。而随着城市功能的转变，当代上海需要滨江绿色生态、休闲空间、宜居和办公环境。还绿于江，还江于民的新阶段终于在2018年元旦得以实现。此课题让夏令营紧扣这一上海城市的脉搏。

而2018年秋开始的本科课程设计，基于黄浦江刚刚贯通的45公里滨江岸线。由于此基地并不在45公里之内，让学生们可以在上海滨水建成环境的基础上，深入滨江感受和调研实际的使用情况，从而对这块即将改造的滨水区域提出理想与现实结合的构思。这种与时俱进的命题，有利于学生在专业学习中理论结合实际。

无论是夏令营成员还是本科学生，应该都从这一"水边的感悟和交流"中受益良多。希望以上体会有助于大家了解此次夏令营及其后续的发展。

As an international summer camp, our purpose is to encourage the communication between Tongji University and other universities no matter at home or abroad. The design theme, "The Revitalization of the Waterfront Portal", is the key point of the summer camp, focusing on the Huangpu riverside.

The Focus on City

The members of the summer camp came to Shanghai from all over the world. They were encouraged to recognize Shanghai as a foothold, and from where they can understand the city, contemplate the city and provide suggestions for the city. Located in the Yangtze Estuary, Shanghai settles its first place in the Yangtze River Economic Belt, holding its great responsibility. The Huangpu River is the mother river of Shanghai and breeds the city in all aspects. At the confluence of the Yangtze River and the Huangpu River, it represents a large number of typical problems during the development of a city, which could be a great point for international students to understand Shanghai from macro to micro.

Learning while doing is the traditional characteristic of Department of Landscape. The undergraduate course combines planning and designing, which provides opportunities for solving practical problems when they are still beginners but on their way to professionals. The design theme encourages students to think about the real problems existing in the representative central areas of Shanghai and to keep their feet on the ground.

The Focus on Profession

Tongji CAUP International Design Summer School is an academic event hosted by Department of Architecture, Department of Urban Planning and Department of Landscape in turn. As a summer camp hosted by Department of Landscape, the design theme introduces a perspective of integrating the universe and humanity. In this area, there are not only abandoned factories and ports needed to be rebuilt and refreshed, but also special hydrology such as confluence of rivers and tidal phenomenon, in which natural water, wetland and wild birds are to be protected. The suggestions and plans made by each team are meaningful answers for realizing the sustainable development of nature and culture in Shanghai.

According to the undergraduate courses, juniors of Department of Landscape are facing the transformation stage of their major study. When they are freshmen and sophomores, the most important thing is to accumulate the basic knowledge of major roundly and diversely. When it comes to juniors, the request for them is to learn landscape architecture more clearly based on the planning and designing of the park, green land system, tourism area and so on. Therefore, the design is suitable for the transformation stage's need, which is asking juniors to balance the ecology and architecture (natural environment and artificial environment).

The Focus on Time

During the summer camp, students started their academic communication at a specific point, 2017. It is the year that witnessed the last stage of connecting the Huangpu riverside. In the view of national industry history, Shanghai is an important city. At that time, numerous ports and piers, factories and storages were built along the Huangpu River. However, as city function has changed, it recalls waterfront ecology, leisure space, livable environment and working environment. The new stage of returning green back to the river and returning the river back to people has finally realized at the beginning of 2018. The design theme makes the summer camp stay close to the present of Shanghai.

The 2018 autumn course is based on the 45-kilometer long Huangpu Riverside. Due to the fact that the site is out of the range, students can explore deeper and experience better the real situation of an established area, then give a rather reasonable design of the site ready to be rebuilt. It is good for students to use their theoretical knowledge more practical when the theme is up to date.

图书在版编目（ＣＩＰ）数据

景观再生：滨水门户的活力复兴 / 同济大学建筑与城市规划学院景观学系著. -- 上海：同济大学出版社, 2020.6
ISBN 978-7-5608-8524-7

Ⅰ. ①景… Ⅱ. ①同… Ⅲ. ①城市－理水(园林)－景观设计 Ⅳ. ①TU986.4

中国版本图书馆CIP数据核字(2020)第123001号

景观再生

滨水门户的活力复兴

同济大学建筑与城市规划学院景观学系　著

出版人：华春荣
责任编辑：晁艳
特约编辑：王玮祎
平面设计：金雅萍
封面设计：张微
责任校对：徐春莲

版　次：2020 年6 月第1 版
印　次：2020 年6 月第1 次印刷
印　刷：上海安枫印务有限公司
开　本：889mm×1194mm 1/32
印　张：4
字　数：108 000
书　号：ISBN 978-7-5608-8524-7
定　价：58.00 元
出版发行：同济大学出版社
地　址：上海市四平路1239 号
邮政编码：200092
网　址：http://www.tongjipress.com.cn
本书若有印装问题，请向本社发行部调换
版权所有 侵权必究